The

MOST ANTICIPATED SPACE IMAGES EVER

The
James Webb
Telescope's First
Cosmic Images

Mary Donald

Copyright©2022 by Mary Donald

All rights reserved. No part of this publication may be reproduced, distributed, or transmitted in any form or by any means, including photocopying, recording, or other electronic or mechanical methods without the prior written permission of the publisher except in the case of brief quotations embodied in critical reviews and certain other non-commercial uses by copyright law.

TABLE OF CONTENTS

INTRODUCTION

CHAPTER ONE

THE JAMES WEBB SPACE TELESCOPE'S FIRST SCIENCE IMAGES

CHAPTER TWO

HOW THE JAMES WEBB SPACE TELESCOPE WILL ACHIEVE UNPRECEDENTED THINGS

CHAPTER THREE

JAMES WEBB SPACE TELESCOPE LAUNCH AND DEPLOYMENT

CHAPTER FOUR

JAMES WEBB SPACE TELESCOPE SCIENCE MANDATE

INTRODUCTION

On Dec 25, 2021, NASA's James Webb Space Telescope (sometimes called JWST or Webb), an infrared space observatory, was launched aboard an Arianespace Ariane 5 rocket.

On July 12, NASA made the first scientific photos from Webb available in a live event. These stunning images will serve as a reminder to the rest of the world of how powerful America is. Nothing is impossible given what we are capable of. The James Webb telescope is a representation of the tenacious spirit of American innovation.

NASA's largest and most powerful space scientific observatory, the $10 billion James Webb Space Telescope, will explore the cosmos to learn more about the evolution of the universe from the Big Bang to the birth of alien planets and beyond. It is one of NASA's Great Observatories, a collection of enormous space instruments involving the Hubble Space Telescope that can look far into space.

The James Webb Space Telescope traveled almost one million miles (1.5 million kilometers) to a Lagrange point, a gravitationally stable place in space, over the course of

30 days. On January 24, 2022, the telescope reached L2, the second sun-Earth Lagrange point. This orbit will enable the telescope to remain in line with Earth as it orbits the sun since L2 is a location in space close to Earth that is opposite from the sun. Other space telescopes like the Herschel Space Telescope and the Planck Space Observatory have found it to be a useful location.

The James Webb Space Telescope will concentrate on four key areas, according to NASA: the origin of the universe's first light, the formation of galaxies in the early cosmos, the emergence of stars and protoplanetary systems, and planets (including the origins of life)

Tens of thousands of scientists are benefiting from this incredible telescope because it will be there for many years; in fact, some of them were just born or haven't even been born yet. Something amazing is waiting to be discovered somewhere. With images of this caliber, science is moving into a new stage of discovery. These highly-anticipated images open up a brand-new view into the evolution of the universe.

Today, the James Webb mission is open for business. This is only the beginning. The best is yet to come. The Webb telescope is designed to make ground-breaking discoveries. With this telescope, it's pretty impossible not to break records.

CHAPTER ONE

The James Webb Space Telescope's First Science Images

For the better part of the last 26 years, NASA scientists working on the James Webb Space Telescope have pleaded for three things: tolerance, time, and, in no small part, money. The idea for a next-generation space telescope to peer 13.6 billion light years away was first put forth in 1996 by a group of astronomers working with the space agency. This telescope would be able to detect infrared light that has been making its way to Earth since only 200 million years after the Big Bang. They asserted that the telescope would be prepared for launch by 2007 and would cost only $500 million—a modest sum by current standards. That's not how it turned out.

The global coronavirus (COVID-19) pandemic hindered JWST's development in 2020, and in July of that year, NASA announced a new launch date of October 31,

2021. The delays persisted despite the JWST team's tenacity and resolve in the face of adversity.

Ariane 5 launch vehicle issues in June 2021 caused the launch to be postponed to November or potentially early December 2021. The observatory had not yet been transported from its original location in California to ESA's launch site at Kourou in French Guiana when NASA and ESA announced yet another delay in September. The two organizations announced a new launch date of December 18, but inclement weather quickly cancelled it.

The launch that was anticipated for November didn't occur until Christmas Day 2021, and what about that $500 million price tag? It eventually reached $10 billion. However, the astronomers' assurance that the new telescope would provide amazing pictures remained unchanged.

But that assurance was shortly fulfilled. A new era of peering into the vast universe has initiated.

The James Webb Space Telescope, the most powerful telescope ever created, has produced its first set of eagerly awaited, full-color scientific images.

The enormous telescope is positioned to peer at some of the oldest galaxies and stars ever created as it orbits the Earth at a distance of about 1 million miles. Since it takes so long for this old light to reach us (or, more precisely, reach the $10 billion Webb telescope), viewing these objects is like looking back into the past by billions of years.

This first collection of ground-breaking images features views of some of the farthest galaxies, a massive star nursery, and enormous cosmic clouds. Additionally, it provides a greater view into a vast planet outside our solar system.

The renowned Hubble Space Telescope, which has captured unrivaled cosmological views for more than three decades, is replaced by the James Webb telescope. The Webb telescope, however, will detect considerably fainter objects and will be able to peer through previously inaccessible clouds of dense cosmic dust because it has a gold-tinted mirror that is over two-and-a-half times larger than Hubble's.

Here are five of the most eagerly awaited space images ever:

SMACS 0723

Sources: NASA, ESA, CSA, And STScI

This image is known as "Webb's First Deep Field" by NASA. "SMACS 0723" galaxy cluster is depicted in the picture. Galaxies that are farther away in the background are distorted and magnified by the mass of the galaxies.

In this image, JWST noticed a population of very distant galaxies. Light is distorted by galaxies in the foreground, which aids in enlarging these distant objects.

Bill Nelson, the administrator of NASA, revealed that the light from those galaxies had been traveling for billions of years. You are specifically viewing the SMACS 0723 galaxy cluster as it looked 4.6 billion years ago. But behind it, there are older galaxies.

According to a statement from NASA, the deepest and cleanest infrared image of the distant universe to date has been captured in this first image from NASA's James Webb Space Telescope. This image of galaxy cluster SMACS 0723, also known as Webb's First Deep Field, is rich in detail.

Webb's view has seen the initial appearance of thousands of galaxies, including the faintest objects ever observed in the infrared. This portion of the huge universe fills a piece of sky about the size of a grain of sand held by a person on the ground with their arm stretched out.

Exoplanet spectrum from WASP-96 b

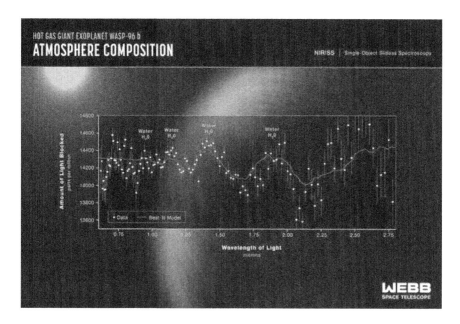

Sources: NASA, ESA, CSA, And STScI

There won't be any nice images to accompany some of the Webb telescope's most breathtaking observations. Webb can investigate the elements that make up the atmospheres of far-off alien worlds using devices called spectrometers. For instance, some planets may have water, methane, and carbon dioxide, which may indicate that they are habitable worlds.

WASP-96 b, also referred to as a "hot Jupiter," provided Webb with the first spectrum of the gases on an exoplanet. It's a high-temperature gas giant that completes one orbit of its star in just 3.4 days thanks to its incredible speed.

The James Webb Space Telescope observed the unambiguous trace of water in the atmosphere surrounding a hot, puffy gas giant planet orbiting a far-off Sun-like star, coupled with proof of clouds and haze. This observation, which deduces the presence of particular gas molecules from miniature brightness changes in specific light colors, is the most comprehensive of its type to date and shows Webb's unmatched capacity to study atmospheres hundreds of light-years away.

The Southern Ring Nebula

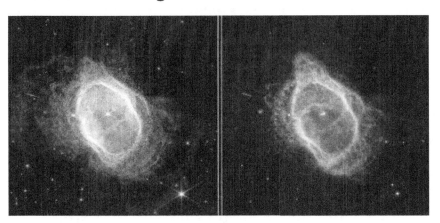

Sources: NASA, ESA, CSA, And STScI

The Southern Ring Nebula is a particular class of object known as a "planetary nebula". These are colorful shells of gas and dust that a dying star has ejected into space. This famous planetary nebula is located about 2,000 light-years away from Earth.

Some stars save their best for last, according to NASA. The James Webb Space Telescope has revealed for the first time that the star in the center of this panorama is shrouded in dust. This star has been spewing rings of gas and dust in all directions for thousands of years.

Stephan's Quintet

Sources: NASA, ESA, CSA, And STScI

A distinguished collection of galaxies located 290 million light-years away is called Stephan's quintet. According to NASA, four of them are "locked in a cosmic dance of repeated close encounters" and are relatively close to one another.

The Webb space telescope shows never-before-seen details in this galaxy group with its powerful, infrared vision and extraordinarily high spatial resolution. The image is adorned with dazzling starburst regions and clusters of millions of new stars. Gravitational interactions cause several of the galaxies to produce sweeping tails of gas, dust, and stars.

The Carina Nebula

Sources: NASA, ESA, CSA, And STScl

Some of the most beautiful parts of space are nebulae. They resemble the enormous dust and gas clouds that result from the explosion of a massive star. They are ideal locations for the emergence of new stars. The enormous Carina Nebula, which is some 7,600 light-years away and where big stars have already formed, was visible to Webb.

The edge of a nearby, young, star-forming region which appears as a landscape of 'mountains' and 'valleys' speckled with glittering stars is known as the NGC 3324 in the Carina Nebula, according to NASA. This image, which was taken in infrared light by NASA's new James Webb Space Telescope, for the first time illuminates previously hidden areas of star birth. The tallest "peaks" you can view are about seven light-years high.

CHAPTER TWO

How The James Webb Space Telescope Will Achieve Unprecedented Things

Webb had already treated us to a number of stunning images during its different instrument checks prior to the first cosmic image and data release.

NASA revealed that the James Webb Space Telescope had taken its first pictures of starlight on February 11, 2022. Webb's first image was of a star known as HD 84406. Webb's 18 mirror segments, which are positioned on the primary mirror, captured light from HD84406 and created a mosaic of 18 dispersed bright dots.

Over the coming months, these 18 dots will gradually merge into a single star as Webb aligns and focuses.

18 unfocused duplicates of a star were purposefully arranged into a hexagonal configuration in a new and enhanced image of HD84406 that was released by NASA on February 18, 2022. The primary mirror's individual

segments will only be successfully aligned by the observatory after which the image stacking procedure will start. This will create a single, clear view with 18 images stacked on top of one another.

An outstanding "selfie" was also captured by Webb using the NIRCam instrument's specialized camera. Engineering and alignment purposes are intended for the camera's use.

One of the mirror segments is shining brighter than the rest in the "selfie" because it was the only one that had been successfully aligned and was pointing at a star at the time. One by one, the remaining mirror segments were correctly aligned.

After proving it can take "crisp, well-focused photos" with all four of its scientific instruments, NASA declared on April 28, 2022 that the James Webb Space Telescope had completed its alignment phase.

How will the James Webb Space Telescope achieve unprecedented things?

Large mirror – Webb's mirror, which reflects light, measures more than 21 feet in diameter. That is more than 2.5 times bigger than the mirror of the Hubble Space Telescope. The ability to capture more light enables Webb to observe further away, older objects. JWST will peer at stars and galaxies that formed about 13 billion years ago, over more than hundred million years after the Big Bang.

The world is going to witness the very first stars and galaxies that ever formed.

Infrared view – Webb is primarily an infrared telescope, which means it observes light in the infrared spectrum, in contrast to Hubble, which primarily views light that is visible to us. This gives us much more access to the cosmos. Since infrared light has longer wavelengths than visible light, it may pass through cosmic clouds with more efficiency; it encounters these tightly packed particles less frequently and is hence less likely to be scattered by them. Eventually, Webb's infrared vision can see through objects that Hubble cannot.

Peering into distant exoplantes – The Webb telescope is equipped with specialized instruments called spectrometers that will completely change how we perceive these far-off planets. Whether they are gas giants or smaller rocky planets, the devices can identify the molecules (such as water, carbon dioxide, and methane) that are present in the atmospheres of distant exoplanets. Exoplanets in the Milky Way galaxy will be studied by Webb.

CHAPTER THREE

James Webb Space Telescope Launch And Deployment

The space telescope was renamed the James Webb Space Telescope in September 2002 after being known as the Next Generation Space Telescope. It was named after James Webb, a former NASA administrator.

From 1961 until 1968, Webb served as the space agency's director before stepping down just a few months before NASA put the first man on the moon.

Webb is regarded as a pioneer in space science, despite the fact that the Apollo moon program is most commonly connected with his time as NASA administrator. Webb established NASA's science goals even during a period of intense political unrest, arguing that the launch of a sizable space telescope should be a top priority for the space agency.

Under James Webb's direction, NASA launched more than 75 space science missions. Missions that looked at the sun, stars, and galaxies as well as the area right above the Earth's atmosphere were among them.

The name chosen for the space telescope has not been well received by everyone. Due to allegations that James Webb participated in discrimination against homosexual and lesbian NASA personnel while serving as administrator, critics started an online petition asking NASA to rename the telescope. In spite of criticism, NASA has since declared that it would not rename the telescope.

NASA reported that the James Webb Space Telescope should have enough fuel to more than twice its minimum mission life expectancy of a decade thanks to a successful and precise launch from ESA's launch site at Kourou in French Guiana. The James Webb Space Telescope has made numerous advances since it was launched.

The observatory was seen flying away from the Ariane 5 rocket that launched it into space in an excellent HD video. Webb slowly drifts away from its rocket stage in the three-minute video as it unfolds its solar panels.

On December 26, 2021, the James Webb Space Telescope deployed and tested a crucial antenna, a procedure that took roughly an hour. The antenna will be in charge of science data dumps to Earth at least twice a day. On December 27, just a day later, the observatory soared outside of the lunar orbit.

Webb successfully extended its enormous sunshield on December 31, 2021. On January 3, 2022, the first of the sunshield's five layers was tensioned. It was finished the following day. On January 5, 2022, the secondary mirror of the telescope was successfully deployed and secured.

The massive primary mirror of the James Webb Space Telescope was finally unfolded and fully deployed on January 8, 2022, according to a NASA announcement. The 18 separate mirrors that constitute the observatory's main mirror must now be aligned. The alignment process took up to 120 days to complete after launch.

On January 24, 2022, the James Webb Space Telescope traveled over a million miles (1.5 million kilometers) to reach L2, the second sun-Earth Lagrange point that it will orbit. The gravitationally stable places in space are called Lagrange points.

CHAPTER FOUR

James Webb Space Telescope Science Mandate

The science mandate of James Webb Space Telescope is primarily focused on four areas:

The first light and reionization

This is the period of time immediately following the Big Bang, when the universe as we know it today first began. In the early moments following the Big Bang, when it was composed of a sea of elementary particles (such as electrons, protons, and neutrons), light was not visible until the cosmos cooled down sufficiently for these particles to start combining.

The "epoch of reionization," which refers to the time when neutral hydrogen was reionized (made to have an electric charge again) by radiation from these earliest stars, is another topic that JWST will investigate.

The ionized hydrogen and helium atoms attracted electrons, converting them into neutral atoms, which for the first time allowed light to travel freely as a result of not scattering off free electrons anymore. The universe wasn't as mysterious anymore. But before the first light sources starts to form and put an end to the cosmic dark ages, it would take some time (perhaps up to a few hundred million years after the Big Bang).

It is unknown exactly how the first stars in the universe (stars that fused existing hydrogen atoms into additional helium) appeared or when they originally formed. These are some of the questions JWST was created to assist us to answer.

Consider the light leaving the earliest stars and galaxies about 13.6 billion years ago and traveling through space and time to arrive at our telescopes. In essence, we are viewing these objects as they were 13.6 billion years ago, when light first left them. We refer to this shift in color or wavelength toward the red as a "redshift" as it occurs by the time the light reaches us. Why? In this instance, it's because Einstein's General Relativity becomes effective when discussing extremely far-off objects. It explains that while the universe expands, galaxies or objects travel farther away from one another as the distance between them widens.

Additionally, any light in that space will stretch, shifting its wavelength to a longer range. Due to the fact that this light reaches to us as infrared light, it can render distant objects very faint (or even undetectable) at visible wavelengths of light.

Redshift is the concept that the light that these early stars and galaxies emit, whether it is visible or ultraviolet, actually shifts to redder wavelengths by the time it reaches our current location. Most of that visible light is typically displaced towards the near- and mid-infrared region of the electromagnetic spectrum for very high redshifts (i.e., the farthest objects from us). Because of this, we require a powerful near- and mid-infrared telescope, just like JWST, to observe the earliest stars and galaxies.

Assembly of galaxies

We can see the large-scale organization of matter in the universe through galaxies. Scientists research how the matter is currently arranged and how that organization has changed across cosmic time in order to comprehend the nature and history of the universe.

In fact, in our search for this knowledge, scientists look at how matter is dispersed and acts at various scales. Each scale provides us with crucial hints about the creation and evolution of the universe, from looking at the building blocks of matter at the subatomic particle level to the vast formations of galaxies and dark matter that span the cosmos.

Magnificent spiral galaxies have been the subject of numerous stunning images captured by telescopes like the Hubble.

However, this is not how galaxies have always appeared. Several diverse events, including the collisions of smaller galaxies, contributed to the formation of the giant spirals we are so accustomed to (including our own). Similar-sized galaxies are thought to collide, disrupt, and merge in order to generate giant elliptical galaxies as well. In fact, since the Universe was six billion years old, it is believed that almost all enormous galaxies have undergone at least one major merger.

In order to understand how matter is arranged on enormous scales, which in turn provides us with clues as to how the universe developed, it is necessary to look at galaxies. One of JWST's objectives is to look back at the oldest galaxies to better understand how the spiral and elliptical galaxies we see now developed from various shapes over billions of years. Additionally, researchers are attempting to understand how galaxies currently originate and come together, as well as how we came to have such a diverse collection of observable galaxies.

We now know that most galaxies have incredibly big black holes living at their centers. What is the correlation between the black holes and the galaxy that hosts them? Additionally, there is much more to learn about the mechanics behind star formation, which might result from a galaxy's internal processes, interactions with other galaxies, or even mergers.

We do know that galaxies are continually forming and coming together today. There are countless instances of galaxies coming together and merging to create new galaxies.

Emergence of stars and protoplanetary systems

Some of the most well-known areas for star formation are the "Pillars of Creation" in the Eagle Nebula. Stars form in gas clouds, and as they get bigger, their radiation pressure drives away the cocooning gas (which may well be utilized again for other stars, if not too broadly distributed). However, looking into the gas is challenging. JWST's infrared eyes will be able to observe heat sources, such as stars developing in these cocoons.

Planets and origins of life

Numerous exoplanets have been found in the previous ten years, including ones using NASA's planet-hunting Kepler Space Telescope. JWST's powerful sensors will be able to take a closer look at these planets, possibly even imaging their atmospheres in certain cases. Scientists would be able to anticipate whether or not a particular planet is habitable, if they have a better understanding of the atmospheres and conditions under which planets form.

Studying exoplanet atmospheres in order to look for the elements of life elsewhere in the cosmos will be one of the key uses of the James Webb Space Telescope. Although JWST is an infrared telescope, how does this help with exoplanet research?

The transit method, which involves looking for diming of the light from a star as its planet passes between us and the star, is one approach Webb will utilize to study exoplanets. This is known as a "transit" in astronomy. We can measure the mass of the planets using ground-based telescope collaboration, via the radial velocity technique (which is measuring the stellar wobble created by a planet's gravitational pull), and JWST will subsequently perform spectroscopy of the planet's atmosphere.

In order to enable direct imaging of exoplanets near bright stars, JWST will also carry coronagraphs. An exoplanet's image would only be a single point, not a broad panorama, but by examining that point, we can discover a lot about the planet. That encompasses its color, seasonal variations, vegetation, rotation, and weather. How does this work? Once more, spectroscopy is the solution.

The science of measuring the intensity of light at various wavelengths is known as spectroscopy. The key to understanding the composition of exoplanet atmospheres is found in the graphical representations of these measurements known as spectra.

When a planet passes in front of a star, the starlight passes through the atmosphere of the planet. If the planet, for instance, has sodium in its atmosphere, the star's spectrum, when combined with the planet's spectrum, will have what is known as an "absorption line" in the region of the spectra where sodium would be expected to be present. The reason is that different molecules and elements absorb light at distinctive energies; and this is how we can predict where in a spectrum we could expect to detect the presence of the signature of sodium (or methane, or water).

Why is an infrared telescope essential for describing these exoplanet atmospheres? It is advantageous to conduct infrared observations because molecules in the exoplanet atmospheres exhibit the greatest diversity of spectral characteristics at infrared wavelengths. Certainly, the ultimate objective is to locate a planet with an atmosphere similar to Earth's.

James Webb Space Telescope really compliments NASA's other solar system missions, such as the observatories on the ground, orbiting the Earth, and in deep space. We can create a more comprehensive image of the objects in our solar system with the use of data from various wavelengths and from many sources, through the aid of Webb's ground-breaking improvements in sensitivity and resolution. JWST will observe Mars, the colossal planets, minor planets like Pluto and Eris, including the smaller worlds in our solar system like comets, asteroids, and Kuiper Belt Objects.

JWST will be used to conduct research that confirm the findings of the Mars rovers and landers and will aid in our understanding of the trace organics in Mars' atmosphere. Webb's observations of the outer solar system will be combined with Cassini's observations of Saturn to improve our understanding of the seasonal weather on our large gas planets. As for asteroids and other tiny bodies in our solar system, JWST will be able to see some features in their spectra that Earth-based observatories can't see. Webb will assist us in learning more about the mineralogy of these rocky objects.

Printed in Great Britain
by Amazon